How To Sync Kindle Books On All Devices

Simple Step-By-Step Guide On How To Quickly Sync Your Kindle And Other Devices

Table Of Contents

- INTRODUCTION 3
- CHAPTER 1 SYNC TO KINDLE 4
  - SYNC OVER THE AIR 4
  - SYNC THROUGH A COMPUTER 7
  - SYNCING ON DEMAND 9
  - TROUBLESHOOTING 10
- CHAPTER 2 SYNC TO ANDROID DEVICES 12
  - SYNCING USING WHISPERSYNC 12
  - SYNCING USING A COMPUTER 13
  - TROUBLESHOOTING 14
- CHAPTER 3 SYNC TO IOS DEVICES 17
  - SYNCING USING WHISPERSYNC 17
  - TROUBLESHOOTING 18
- CHAPTER 4 SYNC TO COMPUTERS (PC, MAC, LINUX) 20
  - SYNCING TO PC 20
  - SYNCING TO MAC 22
  - SYNCING FOR LINUX 23
- CHAPTER 5 THE KINDLE CLOUD READER 25
- CHAPTER 6 PRO SYNC TIPS 28
  - CHANGING YOUR DEFAULT DEVICE 28
  - CHANGE DEVICE NAMES 30
  - SYNC VIA EMAIL 31
- CONCLUSION 32

Introduction

Do you have a Kindle device?

This wonderful piece of tech has awakened the inner bibliophile in millions of people throughout the world. Aside from offering quality reads at affordable prices, the ease and comfort afforded by Amazon's slew of reading apps and devices is unparalleled.

But if there is one problem in the Kindle ecosystem, it's that their sync doesn't always work flawlessly. It's not entirely their fault, as their technology allows for buttery-smooth synchronization – provided everything works right.

This book is meant for those times when there's a glitch, and surely that's a moment many of us have encountered. Regardless of what your platform may be, let this book guide you on how to go about getting access to all your content all the time!

Chapter 1 Sync to Kindle

So you have a new Kindle, whether a replacement or an upgrade of an older generation. The great thing with the Kindle platform is that all of your content is saved in your Amazon account, and can be retrieved at any time.

However, this doesn't happen automatically, and you need to know your way through the process.

There are two ways you can sync your Kindle content from an old to a new device. One is done over the air, via connecting the new Kindle to the Internet so it can retrieve everything from your cloud. The other method is by using a computer as a "bridge" between the old and new devices.

Sync over the air

This is the standard method of syncing, and is supposed to be the easiest method. However, this would only work if you have already previously backed up all your Kindle content from the old Kindle to the cloud. This is usually done as easily as connecting to the WiFi, when the Kindle will automatically download new books and sync changes, annotations, and other information from the book to Amazon's servers for future retrieval.

Syncing over the air takes advantage of Amazon's "Whispersync" feature. This technology not just tracks what books are on one device, it also tracks all changes you're making to it -- right down to what page you're on at the moment. The biggest advantage of Whispersync is that you could drop your Kindle device and resume reading from your phone or computer, and not lose your place in your book.

The only downside is that this system does not recognize "sideloaded" content, or content that have been loaded onto your computer from a PC connection. Whispersync will still load files that have been saved through Amazon's email, or through the "Send to Kindle" email function.

1. Connect your Kindle to your WiFi.

2. After connecting, the Kindle will ask you to have it registered. Log in using the email address and password of your Amazon account. If this does not happen automatically (or if you accidentally dismissed the screen), you can go back to it using Menu > Settings > Registration. Select "Choose to Register using an Existing Amazon Account".

After logging in, you will receive a prompt to sync all your books. Select "All", and wait for the download to proceed.

If you want to selectively sync content from your Amazon Kindle Account to your new Kindle Device, you may "push" the data from Amazon. Here's how it goes:

1. Log into your Amazon account from your browser. Go to Account > Your Orders > Content and Devices.

2. You will see the titles of your books, with an ellipsis ("...") after each. Select the ellipsis of the book you want to Sync, and select "Download & Transfer via USB".

3. You will be prompted with the different devices currently connected to your Amazon account. Select your new Kindle, and click "Download". If your new Kindle is connected to WiFi, then the book should automatically be downloaded.

Sync through a computer

Syncing through a computer is the way to go if, for any reason, your Kindle data remains in the old device without being backed up to the cloud. It is also good for instances when some content remains unsynced, and you are no longer able to connect your old Kindle to the Internet due to damage or the like. This method is also good for those who sideload their own content into their Kindles.

This method is surprisingly easy, and will work so long as you can get the old Kindle to be recognized by the computer. Here are the steps:

1. Plug in your old Kindle via USB cable. The computer will recognize it as a storage, and will show a "Documents" folder. This folder is where all of your Kindle-related data is saved -- from your books, to your annotations, and more.

2. Copy the Documents folder to your computer.

3. Plug in your new Kindle, and copy the Documents folder to it.

4. If you wish, you may also selectively copy ebooks: you may browse the Documents folder you just copied to your computer, and look for the files with the AZW extensions. These are the books you bought from your Amazon Kindle account. Note that some ebooks may have multiple "copies" in your Documents

folder, and all of them will need to be copied. You may also copy various sideloaded content such as PDFs. Make sure among the files copied is the Clippings.txt file, which contains the notes and highlights you made for your sideloaded books.

Syncing on Demand

Kindle devices also have a feature to sync "on demand". Instead of downloading content from the cloud to your device, though, it uploads your most recent activity from your device to the cloud. This is useful as a sort of "refresh" feature. By default, Whispersync only scans for changes and uploads them for every set time interval. By manually syncing, you can make sure that your latest page progress, annotations, and the like are saved to the cloud even before you move from your current Kindle to a different device.

You can sync on demand using the following steps:

1. From the main screen (home) of your Kindle devices, select either the Quick Actions menu (the cog icon in newer Kindle devices) or the Menu option (in older Kindle devices).

2. Select either Sync my Kindle (for newer Kindles) or Sync & Check for Items (for older Kindles). Doing this will activate Whispersync and sync your progress and activity across any other Kindle device and app that you own.

There is also a separate option to just sync the reading progress, available only on older Kindle devices this may be handy if you're on a slow or weak WiFi connection (though newer Kindle devices forego this for a total sync instead). To use this, go to Menu and select Sync to the Furthest Page Read.

Troubleshooting

An important thing to note here is that not all ebooks are compatible with all Kindle models. But, compatibility issues are something to be expected from virtually all platforms. Some Kindle Store content will be marked with a tag that says "Available for the following devices" on the Amazon website. If this tag does not appear, then that means the book is available for all models.

If you're relying on Whispersync to keep your content updated, and the sync fails, you need to make sure that the Whispersync component is enabled. Here are the steps:

1. Go to Settings > Device Options > Personalize Your Kindle > Advanced Options.

2. Check if Whispersync is enabled. If not, select it to turn it on.

3. If the sync still fails, you will also have to check Whispersync under the Settings of your Amazon account. Make sure it is turned on.

If it still fails, you may want to try "refreshing" the Whispersync function by disconnecting your Kindle from WiFi for some time. Here are the methods you could try:

1. Reset your device by holding the power button for 40 seconds, or until the boot-up screen can be seen (the picture of the boy reading underneath a tree).

2. Go to flight mode for 15 seconds, then turn off flight mode.

3. Turn off your WiFi on your device, then go to Settings to reconfigure it.

4. Go to Account Settings and log off then log back in.

If it still fails, contact Amazon's customer support to check if there are hardware issues with the device.

Chapter 2 Sync to Android Devices

Syncing using Whispersync

Android still has the lion's share of the smartphone OS market, and if you enjoy reading via Kindle it's likely you'd also like to take all your collections over the phone. This is useful for when you want to access everything and read even when your Kindle isn't at hand.

Fortunately, Whispersync also covers this aspect really well. In fact, it can be set up in just a couple of minutes:

1. Go to Google Play Store and look for the Kindle app. This is the one with the logo of the boy reading underneath a tree -- same as the Kindle startup screen.

2. Download and launch the app. Sign in using the Amazon account where your Kindle is registered. This will automatically sync your content from the cloud to your phone, along with all your bookmarks and notes.

Note that this has one major limitation. It will never sync the various sideloaded content from your Kindle to your phone. In fact, the Kindle app does not handle any form of sideloaded content automatically, and should you wish to access them on your phone, you need to find another way to sync them (such as through a computer or a different cloud service) and store them locally on your device. You can still use the Kindle app to open them, but they won't be synced at all.

Syncing using a computer

Just like in a Kindle-to-Kindle sync, you have the option of using a computer as an intermediary for loading content from your Kindle to your phone:

1. Plug your phone into your computer via the USB. Make sure you can access its internal storage or SD card, depending on where you have your Kindle data stored.

2. Using your computer, navigate to the folder called "Kindle". This folder houses all the content that your Kindle app uses, and it is also here that Whispersync downloads files when you sync your device.

3. Copy any book or content you wish the Kindle app to access into the Kindle folder in your phone. These content should then be accessible to your Kindle app.

Troubleshooting

Syncing to your Android app has pretty much the same limitations as syncing across different Kindle devices. For one, not every content is available for all platforms, and you need to check if the content you purchase can be used on the Kindle reading app for Android. Also, Whispersync needs to be turned on in your Amazon account, for it to work automatically. To know more about these, check out the troubleshooting section of the preceding chapter.

One other common issue is accidentally using a different account to sign in to the Kindle app, instead of the one where your content is located. To check if this is the case, go to your computer and access Content and Devices in your Amazon account (see Sync over the air in Chapter 1 for the steps). If your phone is not registered here, then you may have used a different account.

If you confirm this to be the case, you can change accounts just by following these simple steps:

1. Launch the Kindle app and click the Menu (hamburger) icon on the upper left corner. If launching the app directly takes you to a book, you can exit by going to Settings > Library. You can find the main Menu option here.

2. Go to Your Account > Settings. From here, tap the Deregister this Kindle option. This will refresh the entire app and ask you to sign in again.

3. Sign in using the same credentials as the account that has the data you wish to sync.

Amazon now implements a "Single Sign On" experience for all its apps, so deregistering your Kindle app could affect any other device or app using the same login information as the one you just signed out from. This is important to note if you are relying on another Amazon account from their ecosystem, signed into the same account as the one you signed out from.

After this, the Android device would now appear as a registered device under the dashboard of your Amazon account online. From here, you can also sync specific content from your Amazon account online to your Kindle app, using the "Download & Transfer via USB" feature detailed in the previous chapter. Again, remember that you will only be finding your Kindle app as a destination device when your content supports reading here.

It is also possible to deregister your device using the web interface. For this, however, you need to log in to the Amazon account that has been mistakenly added to the app. Once you're logged in, follow these steps:

1. Go to Manage my Kindle > Manage my devices. Here, you will see all the devices currently registered on your account.

2. From here, you can see and deregister the app you just connected. After deregistering, access the app on your phone again and you should be "logged out". Log in again using the correct account.

Chapter 3 Sync to iOS Devices

Syncing using Whispersync

Apple devices usually have a superior screen resolution and font management (compared to most other products on the market), therefore they make for really eye-friendly reading devices. It's no wonder many readers are foregoing their Kindle device for a more multi-purpose iPad, or iPhone.

Regardless of your personal reason, reading through your iOS device is a very easy affair thanks to the Kindle app. To get you up and running, download the Kindle app from the Apple App Store. This is the app that has the Kindle logo (boy reading under a tree). Then, once you're done, follow these steps:

1. Launch the app and sign in using the Amazon account where the content you want synced is found.

2. At this point, the app should automatically download all your saved content, including your annotations, bookmarks, and the like.

3. If this does not happen automatically, tap on the menu icon on the upper-left corner of the app. This is the Menu option. Go to Settings > Other. Check if "Whispersync for books" is turned on.

If not, toggle it by tapping. Acknowledge the pop-up by tapping OK.

Just like its Android counterpart, one weakness of the app is that it can only sync files that are loaded through the Amazon web store. That means if you're sideloading files, you will not be able to retrieve them on your iOS device even if Whispersync is turned on across all your devices. As well, there are certain ebooks that will only work for certain devices, so you might want to double-check if the books you want to read are marked as compatible for the iOS Kindle app.

Troubleshooting

Aside from the similarities in limitations, the Android and iOS Kindle apps are also similar in their troubleshooting methods.

The most common reason for items not syncing will still be a mistake in the login credentials provided upon the first launch of the app. A different login might have been added. You can check this by logging into your Amazon account through a computer browser. Go to Account > Your Orders > Content and Devices. You should see the newly-registered iOS device linked here. If that is not the case (and if you are sure you have a working WiFi connection), then it's likely you're signed in using a different account.

To resolve this, you need to "log out" by deregistering the Kindle app from your Amazon account. This is how it's done on iOS:

1. Launch the Kindle app, and go to the Home screen of the app by tapping the Home button.

2. On the lower right-hand corner, tap the Info icon (the small "i").

3. At the top of the screen, you will find "Registered to...". Tap this, and select Deregister. If you are using an iPad, these same options can be found under the Settings option.

4. Now that you have deregistered, the Kindle app is essentially refreshed. Log into it again using the correct email and password. As long as Whispersync is turned on, you should be able to retrieve your ebook data.

You can also deregister using the online interface, by following the steps under the Troubleshooting section of the preceding chapter.

Chapter 4 Sync to Computers (PC, Mac, Linux)

Syncing to PC

Sometimes you would just want to stay put while reading your favorite book. Or you want to consume your content on a bigger screen. Whatever your reason, you would be happy that all your Kindle content can also be synced to your desktop computer!

Let's start with desktop PCs, running the Windows platform. All you have to do is download the latest version of the Kindle app either from the Amazon website, or from the Windows app store (if you are using Windows 8 or above). Simply log into the app, and all your books will auto-populate!

It's that easy, and the only reason the sync won't work correctly is if you don't have the correct Amazon account logged in, or if Internet connection doesn't work correctly. Just like Kindle devices and mobile apps, logging into the Kindle PC app will register your device on your Amazon cloud account. If you go to your Amazon account, and then to Account > Your Orders > Content and Devices and do not see the PC registered, you would need to deregister your PC app and sign in using the right account.

To deregister, you can follow these steps:

1. Click on Tools on the toolbar at the top of the Windows Kindle app, then go to Options.

2. On the left of the popup window, click on Registration, and you will see your Kindle's registration info. Click on Deregister, and Save.

Once you deregister, your Windows Kindle app will return to its original state, and you need to log in again. Use the email address and password connected to the account containing the items you wish to sync, and the files will automatically download.

It is also possible to deregister your computer through Amazon's online interface:

1. Sign into your Amazon account, and go to the Manage Your Kindle page.

2. From here, scroll down to the section containing Your Computer and Other Mobile Devices. You will see a list of devices currently registered. Find your PC, and click on the Deregister link at the right. After this, you will be asked to confirm your action.

Syncing to Mac

Kindle has also created an app for Mac users, which can be found in the Mac App Store. All you have to do is enter the words "Kindle App" in the search box, download the app, open it, and click Install. The interface and usage of both the Kindle apps for PC and Mac are pretty much the same, and even the troubleshooting steps are similar. Kindle has made sure that users of its apps get a uniform experience in as much as the platform allows it.

Once you download the Mac version of the Kindle app and successfully installed it, log in using the same Amazon account as the one that has your Kindle ebooks. Like in Windows, the app would automatically be downloaded to your device.

The limitations are also pretty much the same. The Mac version of the Kindle app also does not allow for the syncing of sideloaded apps, so all you get will be the files officially bought and/or downloaded from the Kindle Store.

Syncing for Linux

Linux is a special case, as while its share among desktop users is inching higher as time passes, the overall number is still too low to warrant Amazon to take time to create a dedicated app. The average Linux user can only automatically sync his ebooks and other Kindle Store content through workarounds.

For the Kindle user on Linux, the app called WINE is a great friend. This is a software that allows Linux users to install apps from the Windows platform. This means you can download the official Windows for PC app, and the WINE system will have it up and running. However, as the Kindle app evolves internally with each new version release, WINE might not always be good enough to run its latest programming. Therefore, the best way to run the Kindle App through this method is through downloading an older extension (usually though an external service or file mirror that hosts it). These older versions may not have of the latest snazzy features Amazon Kindle has for its users, but it's a great way for those using Linux to join in the party.

Another option is also available, though it is more time-consuming. However, this method is great as it allows you to store offline copies of your Kindle Books, ensuring you don't lose them to the cloud for any reason.

This last method involves using an online interface to read your content, which will be detailed in the next chapter. While this is not an "app" per se, this Cloud Reader ensures that you can always access your ebooks so long as you have a working browser, input device, and a monitor.

Chapter 5 The Kindle Cloud Reader

If you are having difficulties with syncing, or if the book or content you wish to read does not support the Kindle app for your computer, you can always use Amazon's own Cloud Reader! This is also good if you are using a public computer, or a borrowed one. The Cloud Reader is an in-browser reader that acts as a reader app for your Kindle content. There is also a Chrome app available, which accomplishes the same function.

Unlike the app, though, there are some very specific settings your browser should have so you can use it. The following browsers are officially supported:

- Google Chrome v.20 or higher
- Mozilla Firefox v.10 or higher
- Safari v.5 or higher
- Internet Explorer 10 or higher

Kindle Cloud Reader can be accessed at read.amazon.com. You will still need the email address and password of the Kindle account that has the content you wish to read. From here, you will be redirected to the option where you can read your content.

One of the things you can do with the Cloud Reader is to enable Offline Mode, which is great for keeping in sync with your readings even when you go offline. Offline Mode works by downloading your content and storing it once you go online, so you could read when you're not. This is useful for staying in sync especially when you are on the road.

You can activate Offline Mode by downloading your Kindle ebooks. You will find a Downloaded tab right beside the Cloud tab at the top of your screen. You can right-click on a book, and click on Download & Pin Book to start the process. The browser app will prompt you to enter offline mode, but it can be a bit spotty if you are using the read.amazon.com version. A better way to go offline would be to download the official extension through the Google Web Store.

One thing you need to remember is that the Cloud Reader is browser-specific. This means you will not be able to retrieve your books from another browser in your computer. You may choose to download a book on one browser, and another in a different one, if you so wish -- they will remain on different spheres and will not mix.

Another function of the Cloud Reader allows you to access your notes and highlights. This is saved in the Notebook format, and can be accessed through the rectangular icon beside the Kindle logo on the homescreen of the browser app. Clicking this button will open up a new tab, which will in turn open up a different tab beside the other ones.

That said, the Cloud Reader has a few limitations:

- Collections are not available, and you cannot sort your ebooks. While this is available on mobile apps, the Reader will download your content individually. Your collections are still saved as they are in the cloud, though.
- Non-Kindle store files are not available. Even if your account is connected to the Amazon cloud, you will not be able to access files that did not come from Kindle. This includes items you send through the Send via Email service. They are still in your library, but are not accessible via the Cloud Reader.

Note that if you are trying to find the Kindle Cloud Reader through various settings, you may also encounter it as Amazon Cloud Reader.

Chapter 6 Pro Sync Tips

Changing your Default Device

One of the main things that can cause confusion in syncing your devices is the fact that Amazon's system of listing your different devices can be a little out of whack. If you're like the average bibliophile with several devices linked to the same Kindle Account, you would see a long list when going to Your Account > Manage Content and Devices. This would look like:

Kindle 1
Kindle 2
Android Device 1
iOS Device 1
PC 1
PC 3

This list is further made confusing by the fact that one of these devices is the "default device", or the device that will first receive the purchased book or free sample. For example, you might be carrying Kindle 2 and accidentally delivered the content to Kindle 1. You would have to wait for Kindle 1's Whispersync to kick in before you can receive it on Kindle 2 (whose own Whispersync also has to run).

Of course, there is the possibility of choosing the receiving device every time you purchase something. Or, better yet, you can change your default device into your go-to device, so you are most likely to receive your content anytime.

By default, the oldest device added is the default device. So if you had a Gen 3 Kindle and first added it years ago, Amazon would continue to deliver to it first. To change that, follow these steps:

1. Launch your browser, and log into your Amazon Account. Go to Your Account > Manage your content and devices.

2. Go to the Your Devices tab in the middle. This will give you a list of all the devices and apps you have associated with your account. Click on the one which you want to set as your default.

3. The next popup will depend on which device you had chosen. If you chose an app for the PC, or an old-generation Kindle devices, simply select Set as default device, a link located at the bottom left. If you are using a more modern product such as the Fire, Fire Tablet, Fire Phone, and the like, the option you want is hidden beneath Device Actions. Select that, and choose Set as default device. This will give you a confirmation popup, where you have to click Save.

Change device names

One difficulty you may find with the steps above is that you would first need to figure out which Kindle device or app is which. We already know that the order shown is from the oldest to the newest device, but you may not be sure which device came before which.

Fortunately, Amazon has provided a way to change the device names of each device. If you can track which one is which, you can now enter identifying words to avoid confusion. Here's how:

1. Go to Account > Manage your content and devices. Go to the list of devices, and click on the device that you wish to rename.

2. Click the Edit link beside the name, and you will be taken to a popup where you can change the display name of the devices. You may also change the email address connected with the same account.

3. Depending on the device, you can also go into the e-reader itself and look for the Settings menu. Here you will find a field where you can edit the device name.

Sync via email

We have previously mentioned the Send via Email feature, and it is also a useful alternative if you want to sync a non-Kindle piece of content to your Kindle devices or apps. You can send a variety of document formats, from the standard ebook and text formats to a full HTML webpage file, and have them all open by Kindle.

To do this, simply attach the doc to an email and send it to your provided forwarding email address. You can find your personal email address by going to the Settings option, and checking the Send to Kindle Email Address. If you want a truly native experience, you can use the subject line "convert" and Amazon would convert it to a Kindle format document with all the latter's advantages! In the Kindle app, these sideloaded content will appear under the Docs folder.

Conclusion

Wasn't that easy? Thanks to the ecosystem Amazon has built, it doesn't take much troubleshooting for your sync to work properly. There are also lots of similarities between the instances of the Kindle app even across different platforms, so you get the same experience every time – it's not hard to get used to usage and troubleshooting, too!

So grab your Kindle, and enjoy reading wherever you are. Get all your books, and all your bookmarks, notes, and settings without breaking a sweat. After all, reading was made to break barriers, and not even tech glitches and fuzzy sync processes should stop that!

I wish you the best of luck!

To your success,

William Seals

www.ingramcontent.com/pod-product-compliance
Lightning Source LLC
Chambersburg PA
CBHW031507210526
45463CB00003B/1118